Thank you for sharing the Garden's vision and
generously supporting the
Growing a Legacy for Generations
Capital Development program.

Bill Hwizuigh

OASIS IN THE CITY

The History of the Desert Botanical Garden

by Tara A. Blanc

HERITAGE PUBLISHERS, INC.

Oasis in the City
The History of the Desert Botanical Garden

Written by Tara A. Blanc
Edited by Susan Zeloznicki
Designed by Gopa Design and Illustration

Published by
Heritage Publishers, Inc.
5312 N. Twelfth Street, Suite 302
Phoenix, Arizona 85014-2903
(602) 277-4780 (800) 972-8507

ISBN 0-929690-51-6
Library of Congress Control Number 00-132383

Printed and bound in the United States of America

Dedication

Had it not been for the devotion and persistence of Gertrude Divine

Webster, the Desert Botanical Garden might not exist. This book is

dedicated to Mrs. Webster and all of the volunteers and staff members

who over the years have followed her example in offering their time and

talent to make the Garden a very special place.

Contents

Acknowledgments

THIS BOOK WAS WRITTEN to record as accurately as possible the people and events that have shaped the Desert Botanical Garden. The many people who were involved in developing this project deserve recognition for their contributions.

Dianne Bean, Dr. William Huizingh, Kayla Kolar, Carolyn Polson O'Malley, Carol Schatt, and Dr. Liz Slauson served as members of the book project committee.

Several Garden members and present and former staff graciously shared their memories through personal interviews. A written account of the Garden's history by the late Elizabeth Fritz was of invaluable assistance.

To all of the people who shared their time and recollections, thank you. This book exists because of you.

Foreword

IMAGINE, IN TODAY'S INSTANT ACCESS WORLD, a handful of people dedicating themselves to planting a garden they would not live to see mature. That's what the founding members of the Arizona Cactus and Native Flora Society did in 1939, sowing the seeds for an institution that would one day become a world renowned center for the study and conservation of desert plants.

Today's Desert Botanical Garden owes its success and stature to the people who over the past sixty years devoted their time, talent, and energy to building and nurturing it. True to the mission outlined in 1939, they helped shape the Garden not just for themselves, but for generations to come.

This is the story of the Desert Botanical Garden. Its history is as rich and colorful as the collections it preserves. Those of us who have been a part of it have been greatly enriched by our experiences here. For those whose experience with the Garden is yet to come, we hope this history will serve as a foundation for the Garden they will create.

Carolyn Polson O'Malley, Executive Director
The Dr. Willliam Huizingh Endowed Chair, Desert Botanical Garden

Gertrude Webster's Camelback Mountain estate was surrounded by extensive gardens, including an impressive array of desert plants.

CHAPTER ONE

PLANTING THE SEEDS

P HOENIX, ARIZONA, in the early 1930s was a bustling, growing community of more than fifty thousand people. There was land as far as the eye could see, but it already was being built up, plowed under, and drilled with prospectors' holes. Few of those early go-getters saw value in the desert plant life that was being destroyed in pursuit of the American Dream.

One person, however, was worried about the way the desert was being ravaged. Gustaf Starck, who was born in Sweden in 1871, had become fascinated with desert plants as a boy. Starck's father, a captain in the Swedish Royal Navy, fueled his son's interest by bringing home exotic specimens collected during his travels. An engineer by trade, Gustaf Starck immigrated to the United States around the turn of the century and worked in the Midwest until moving to Arizona with his wife and children in 1919. Over the years, he developed love and concern for the desert. He bought land in the little country town of Scottsdale, where he planted citrus groves and created a showplace of cacti and succulents.

Starck worked as an engineer with the Salt River Valley Water Users' Association, which provided him a state permit to collect plants. He researched the plants in his collections in books and journals or, when those failed to provide answers, he prepared herbarium sheets to send to the U.S. National Herbarium in Washington, D.C., for identification. As a result of his interest and diligence, he became the local authority on cacti and succulents.

Fellow plant enthusiasts began to trek over dirt roads from Phoenix to the Starck ranch in Scottsdale to learn more about desert horticulture. They were guided by Starck's handmade signs pointing the way and proclaiming "Save the Desert." A study group soon became a regular Sunday event, with various members hosting meetings in their homes. Led by Starck, the group was concerned about the demise of the desert and began to look for a way to preserve some of the Valley's native landscape.

Without the focus, drive, and repeated support of Gertrude Divine Webster, the Desert Botanical Garden might never have been.

Save the desert

On April 18, 1934, a dedicated group of sixteen men and women organized the Arizona Cactus and Native Flora Society. The Society's aim was to save a part of the desert by creating a botanical garden. The garden would display and interpret the desert to those who failed to appreciate its beauty. The Society also aspired to create a living laboratory of international importance.

Unfortunately, many Phoenicians were not interested in such a concept and balked at supporting it. After all, why worry about digging up the desert when there was so much of it? Many of the newcomers from the north and east preferred to import their grass lawns and mulberry trees rather than landscape their yards with the native flora. Besides, when there was economic gain to be made from the land, who would care about a few cacti being bulldozed out of the way?

Starck wrote impassioned letters to the editors of the local papers, gave eloquent speeches at public gatherings, and did everything he could to promote his cause. He won the support of the Phoenix Chamber of Commerce and was appointed to its parks and improvement committee. But it would take a more authoritative and aggressive figure to transform the dream of a botanical garden into reality.

Not to destroy, but to glorify

Gertrude Divine Webster was a woman of means, enjoying considerable wealth from a family lumbering business. She arrived in Phoenix as a winter resident in 1928 from Vermont, building an adobe mansion on eleven acres on the south slope of Camelback Mountain. She returned from a trip to Switzerland in the early 1930s with some rare cactus plants and was referred to Gustaf Starck for advice on raising them. Starck cultivated Webster's interest in cacti by inviting her to join the new Arizona Cactus and Native Flora Society. In 1936 he nominated her as president.

Webster wasted no time in garnering community support. She persuaded her rich and influential friends to host benefits for the proposed botanical garden, while her own home was the scene of many luncheons and dinners where guests with important business and political connections heard speeches on Sonoran Desert plant life. The owners of *The Arizona Republic* and *The Phoenix Gazette* and radio station KOY were named to the project's Advisory Board. Generous with newsprint space and air time, the media helped the Society raise funds and negotiate for the site it had selected as ideal for its purposes—several acres of Papago Park.

Once a town-site for the Pima and Maricopa tribes, the area had been dedicated as a national monument in 1914 because of its excellent specimens of saguaro cactus (*Carnegiea gigantea*) and other desert flora. Named the "Papago Saguaro Monument," it survived until 1930 when the saguaro population was so decimated by the city's growth that the national monument status was transferred to Tucson. The land was granted to the state and administered by the Department of Game and Fish.

The site was ideal for a desert botanical garden. Its terrain was both rolling

Gustaf Starck's concern about the desert's demise led to the founding of the Desert Botanical Garden.

and flat with light and shade. It offered a place where plants from desert climates around the world could flourish.

On November 22, 1936, the 121 members of the Society met at the proposed site to make plans for their garden. They adopted a slogan written by Gertrude Webster, "Not to destroy but to glorify," as their theme. But their noble aims and ambitious plans did not impress the Arizona State Legislature, which carried the authority to lease Papago Park land. The Society's initial petition to lease the land

Gertrude Webster offered the services of her personal gardener, Arthur Johnson, to help with the first planting in the Garden in 1939.

was rejected because it contained a clause providing $2,500 a year for maintenance at the state's expense. The Society appointed a committee to shepherd a second petition—without the offending clause—through what Webster described as the "byways and subways" of the Arizona Capitol.

While they waited, the Society's members continued to raise funds and seek friends for their cause. The Society held its first Cactus Show in April 1938 at Tropical Groves Nursery on Thomas Road and Chicago Avenue (now 40th Street), which attracted many locals, including one thousand school children.

Undaunted by the specter of defeat, Webster continued to issue instructions from her summer home in Vermont for the garden's layout. She helped raise $40,000, donating one-fourth of that amount herself, so the Society could be ready "to demonstrate all that Arizona has failed to recognize — a great natural legacy."

At long last, land

Webster's confidence and the Society's perseverance finally paid off. On July 1, 1938, the State Land Department and the Department of Game and Fish granted a permit for the Society to occupy "certain lands in former Papago Saguaro National Monument for an indefinite period." The agreement stipulated that the Society must begin the propagation and culture of cacti and native flora within six months or lose the rights to lease the land.

Charles Gibbs Adams, a Los Angeles landscape architect who had designed the estates of William Randolph Hearst and Cecil B. DeMille, drew up the preliminary plans for the garden. He noted in the February 1939 issue of the *Cactus & Succulent Journal* that the general plan for the first sixty acres was for "native cacti of Arizona, planted in low drifts as nature does it, to look as though the hand of man had never touched them." The plan was so successful that early visitors to the garden would ask if anything had been planted yet.

On December 12, 1938, the first cacti were planted in the Desert Botanical

Garden on a knoll with a southwest exposure. It was named Kreigbaum Knoll for L.L. Kreigbaum, an agriculture teacher at Phoenix Union High School and a hard-working board member who donated 450 barrel, saguaro, and hedgehog cacti from his own land.

Gustaf Starck contributed his collection of seven hundred specimens. H.O. Bullard, a winter resident from New Jersey, offered a rare collection of agaves from California. Webster donated hundreds of plants from her estate and the services of her personal gardener, Arthur Johnson. Phelps Dodge gave hundreds of

salvaged cuttings of organ pipe and barrel cacti threatened by its open-pit mining operations near Ajo, Arizona. Society members hauled in the cuttings, along with several Joshua trees from the Congress Junction area.

With the land finally acquired and planting underway, the Desert Botanical Garden was taking shape. Gustaf Starck's dream had come true, and it was time to share it with the public.

These barrel cacti from Ajo, Arizona were part of the first planting effort in 1939.

This boojum tree was among the original specimens
Dr. George Lindsay brought to the Garden from Baja
California. It died and was replaced some years later.

CHAPTER TWO

THE FIRST BLOOM

N FEBRUARY 12, 1939, the years of effort and perseverance by Gustaf Starck, Gertrude Webster, and the members of the Arizona Cactus and Native Flora Society culminated in a ceremony held a quarter-mile north of Hole-in-the-Rock in Papago Park. The event was the official dedication for the Desert Botanical Garden. More than two hundred people showed up to help mark the occasion.

Among the notables who attended were Arizona Gov. Robert T. Jones and Phoenix Mayor Walter Thalheimer. Harry M. Fennemore, a Phoenix attorney who had been instrumental in bringing about the agreement between the Society and the State Land Board, presided over the ceremony. Dr. Forrest Shreve, director of the Carnegie Desert Laboratory, offered observations on "Dwellers of the Desert." Gertrude Webster also spoke to the crowd.

"Our purpose is three fold," Webster noted. "We wish to conserve our Arizona desert flora, fast being destroyed. We wish to establish scientific plantings for students and botanists. We wish to make a compelling attraction for winter visitors."

By the time the Desert Botanical Garden was dedicated, plans for the Garden's growth and development were well under way. Annual membership fees were established. Designs, drawings, and a budget had been devised for the first building, a combined assembly hall, herbarium, and library. Only one thing remained to be done — hire a full-time director.

Lindsay takes the lead

In the spring of 1939, George Lindsay was a twenty-two-year-old botany student who had completed his junior year at San Diego State College. During the course of his studies, he had submitted several articles to Scott Haselton, editor of the *Cactus and Succulent Journal*. Lindsay was waiting to transfer to Stanford University

Dr. George Lindsay was the Garden's first director. He served one year, from May 1939 to May 1940.

when he received a visit from Haselton, who had a very interesting offer.

"Scott had recommended to Mrs. Gertrude D. Webster that I be considered for the directorship of the Garden," Lindsay wrote in 1978. "It was an exciting prospect. Dr. Robert T. Craig and I were planning a plant-hunting pack trip to the Barranca de Cobre, Chihuahua, and Mrs. Webster asked me to come by Phoenix for an interview on the way."

Lindsay drove his Model "A" Ford, loaded with plant presses and camping gear, to Phoenix. He met with Webster at her Camelback Mountain estate, and together they toured the Garden's site in Webster's black limousine with her Pekingese dogs. After a pleasant visit, Lindsay headed to Mexico for four weeks of plant collecting before returning to San Diego.

"I talked to Bob Craig about what I'd seen, but I really didn't take it very seriously," Lindsay said in a 1994 interview. "When I got back to San Diego, there were two or three urgent letters from Mrs. Webster. It was contagious, and for a youngster, it was a great opportunity."

Mrs. Webster was so persuasive that Lindsay reported to the Desert Botanical Garden for duty on May 1, 1939, at a salary of $100 a month.

By the time Lindsay took on the directorship, a two-inch water main had been brought into the Garden, the acreage had been surveyed, and preliminary planting was under way. Lou Ella Archer, a member of the Garden's first Advisory Board, arranged to have a crew of twenty-five National Youth Administration (NYA) boys assigned to help the new director fence the Garden's approximately three hundred acres with barbed wire and install thousands of feet of drain pipe. They hauled in several hundred truckloads of granite boulders from Camelback Mountain and topsoil from Paradise Valley to construct and fill plant beds, and edged the Garden's path with decorative rock.

Lindsay spent much of his time in the field. One of his chief collecting partners was Herbert Bool who, with his wife, Angela, ran the Sandyland Cactus Gardens on Camelback Road. The Bools had been founding members of the Desert Botanical Garden and Angela Bool served many years as secretary of the Executive Board. Lindsay and Herb Bool collected in northern Arizona and along the Mexican border during the Garden's first summer. A trip halfway down Baja California to find rare cacti took Lindsay into country where the road tracks were so narrow his truck's double rear tires could not pass. The only way he could get to the specimens he wanted was to remove the outer tires. He also made several trips to California to bring back contributions from the Huntington Botanical Gardens and several commercial growers.

Webster Auditorium under construction in 1939.

The first plantings around Webster Auditorium were completed in time for the dedication in 1940.

Charles Fleming, the Garden's second director, served from 1940 until 1941.

The Garden was taking shape under its young director, including the biggest project of all—the construction of an administration building.

The Webster Auditorium

The Arizona Cactus and Native Flora Society discovered quickly that planning to build a building was one thing, but finding the funds to do so was another matter entirely. The building committee first approached the Works Progress Administration (WPA), which agreed to erect the building for $16,000. But the WPA demanded sponsorship by the State Land Board and State Land Commissioner, who in turn decreed that they would have no financial responsibility. When it became evident that WPA funds would not be available for several months and that there would be a host of restrictions placed on them, the building committee decided to start over.

Bids were obtained from local contractors, with the lowest coming in at $13,000 from Broman and Chapman Construction Company. The Society had only $8,000 in its coffers, so Gertrude Webster agreed to underwrite the balance. It was not an outright gift; instead, each member was to participate in a state-wide drive for funds to reimburse Webster.

The Society broke ground for the building in June 1939 on a knoll that offered a commanding view of the mountains and the Salt River Valley. The pueblo-style structure included a large meeting room and library, a president's office, and two guest rooms at the east end of the building for visiting scientists; and an apartment for the director at the west end. Its adobe walls were up to thirty inches thick, with log vigas brought in from Flagstaff spanning the main hall. A landscape painting above the fireplace was a gift from Phoenix artist Oscar Strobel, who also decorated the upper walls with a symbolic frieze of Indian figures.

The building was completed in October 1939. Furnishing and landscaping the building were the next challenges. Many of the furnishings were donated, including some museum-quality antique pieces. Others had to be purchased, such as the tin chandeliers, candlesticks, and brightly painted chairs bought in Mexico. Some furniture, glassware, china, and silver—necessary for proper entertaining—were purchased on time payments. Money was so tight, however, that the Dorris-Heyman Furniture Company began dunning treasurer Ed Walker for payment. Gertrude Webster was forced to pay the bills herself to avoid a lawsuit.

As dedication day drew closer, crews worked feverishly to complete the landscaping around the building. The specimens from Lindsay's trip to Baja were planted to the immediate north of the building. Aloes lined the east wall, blooming for dedication day, and the patio to the south displayed a variety of night-blooming cereus.

Finally, on January 21, 1940, the new building was dedicated as Webster Auditorium. Nearly two thousand people gathered for the event, where they heard

speeches from the governor, the mayor of Phoenix, and several civic dignitaries, including Gertrude Webster.

"We are building for this state now and for future generations," she told the crowd. "You and I may not be here to see these gardens mature—but perhaps we can look down from heaven and enjoy them just the same."

With the dedication of the auditorium, the Garden's basic infrastructure was in place. More than five thousand specimens had been planted in the Garden and, under Lindsay's direction, standards for collecting data on those plants had been established. But Lindsay was ready to move on and he resigned on June 1, 1940. Coupled with world events, his departure would signal some dark years ahead for the fledgling Desert Botanical Garden.

Changes and challenges

When Lindsay was hired as director, he had told Gertrude Webster that he planned on being at the Garden for only a year. At the time of his resignation, Webster offered him more money, but Lindsay decided it was time to go back to San Diego. He served in the Army in World War II, completed his studies at Stanford University, and later became director of the California Academy of Sciences in San Francisco.

"The year spent starting the Desert Botanical Garden was a wonderful experience for me. However, my personal agricultural interests in California and desire to complete my formal education led to my resignation," Lindsay noted in his 1978 memoir. "Mrs. Webster was disappointed and hurt; then relieved to find a successor in Mr. Fleming."

Nearly two thousand people attended the dedication of Webster Auditorium on January 21, 1940.

19

Taken from the roof of Webster Auditorium in 1940, this view shows nearly all of the Garden that was planted at the time.

"Mr. Fleming" was Charles Fleming, Jr., a native Arizonan with a master's degree in ecological botany. A high school science teacher in Clifton, Arizona, he had just one hour's briefing from Lindsay before assuming his duties as director. Fleming's wife, Ardath, was appointed as "acting hostess." They moved into the director's apartment in Webster Auditorium at a combined monthly salary of $150.

The Flemings nurtured the Garden through its second year, fighting the jackrabbits that ravaged the plants on a daily basis. With the help of eight NYA boys, Charles Fleming hauled in leaf mold and sand to build cactus beds, and built entrance gates of rock on the north (McDowell Road) and south (Van Buren Street) sides of the Garden. He conducted classes for children and adults on desert plants, often leading the adults on field trips. Every Sunday there was a public lecture in the auditorium featuring guest botanists such as Lyman Benson. Fleming was a frequent speaker at Garden Club meetings around the Valley to promote the Society's cause.

Despite the Flemings' efforts, however, things were not all well with the Garden. The Society's finances were strained to the breaking point, with no money for a lath or propagation house, equipment, or library material. Gertrude Webster, suffering from ill health, declared that she could no longer carry the financial burden. The occasional rental of the administration building for group functions and dues from the 138 members of the Society provided the only income.

It was a bleak scene on January 20, 1941, when the Society's Executive Board gathered in Harry Fennemore's office for an emergency meeting. Three possibilities for sharing control and expenses were discussed: ask the state to take on the Garden as an educational project for the University of Arizona; ask the City of Phoenix to adopt the Garden as part of its park system; or ask Arizona State Teacher's College at Tempe (ASTC), now Arizona State University, for some type of cooperative agreement.

The third option proved to be the most viable, and on March twenty-first the Society's Executive Board and representatives from ASTC drew up an agreement. The college would provide for the care and maintenance of the Garden, eliminating the expense of a director, while the Society's members would take responsibility for an educational program. By this time, Fleming, who was not included in the board meetings, had grown weary of the twenty-four-hour-a-day, seven-day-a-week grind, and resigned after one year as director, coincident with the new agreement. He went on to a posi-

Director Charles Fleming built these entrance gates in 1941 with the help of a National Youth Administration crew.

21

This photo of Webster Auditorium was taken by director Charles Fleming in 1941.

tion as an instructor at Riverside Military Academy in Georgia, and later retired to Arizona.

The cooperative plan with ASTC was launched in the summer of 1941 with Gilbert Cady, the college's business manager, serving as supervisor. A college maintenance employee, Bob Svob, moved into the director's apartment with his wife to serve as caretaker. But the United States' entry into World War II brought the cooperative agreement to a halt. Cady was drafted, and several young men who followed Svob as caretaker left in quick succession for better-paying jobs in the defense industry. An elderly couple named Coldwell eventually moved to the Garden

as caretakers, and in 1942 the board decided to close the administration building, except for the occasional Sunday lecture. The few winter visitors who did find their way to the Garden were shown around by the caretaker.

Taking care of the Garden was too much for the caretakers without support, and neglect set in. Rabbits were again on the rampage, as were unscrupulous people who helped themselves to many of the Garden's plants. Military units stationed in Papago Park dragged field guns through the Garden's beds and even shelled the area for target practice. Wartime gasoline rationing brought Garden activities to a halt, including Society meetings. Illness as well as the wartime restrictions on transportation kept Gertrude Webster in New York, and without her on the scene to prod and push, interest waned in keeping the Garden going.

A few faithful members, however, continued to make the long trek out to the Garden on occasion to water the plants. Ed and Katherine Walker were among the few who stuck with it, sending reports to Gertrude Webster about the problems at the site and the appalling lack of local interest in the Garden.

Thanks to the loyal efforts of people like the Walkers, important pieces of the Garden managed to survive. Once the war was over, there would still be the beginnings of a scientific collection, a basic infrastructure, and many plants still growing in the Garden.

Signs point the way to the Garden in the 1940s.
Access to the Garden became easier when Galvin Parkway was constructed in 1963.

23

Director W. Taylor Marshall led the Phoenix Advertising Club on a tour of the Garden in the late 1940s.

CHAPTER THREE

A GARDEN TAKES ROOT

B Y THE END OF THE WAR in 1945, the Desert Botanical Garden was gasping for breath. Much of the collection was missing or damaged, membership had dropped to nineteen, and there was little left in the treasury. The few remaining members set about renewing public interest in the Garden. Quick action was needed to bring it back to life.

Sadly, Gustaf Starck was not among them. He had withdrawn from participation in the Society in recent years due to failing health and frustration over what was happening at the Garden. Starck died in 1945, before his beloved Garden reopened.

The first order of business for the Garden's "survivors" was to hire a new director. George Lindsay, back from military service, was asked by Gertrude Webster to return. He declined the invitation and instead recommended William Taylor Marshall for the job.

A salesman by profession, Marshall was a self-taught botanist who had served

as president of the Cactus and Succulent Society of America in 1938. His interest in botany was sparked on his thirtieth birthday when his daughter gave him a dime-store package of five plants, none of which he could identify. In the ensuing years, while he worked as a sales manager for a cigar manufacturer and then as the owner of a Los Angeles interior decorating firm, Marshall pursued his botanical studies. His interest drove him to make time for field trips and publish several works on cacti and succulents.

The directorship of the Desert Botanical Garden offered Marshall a way to combine his business experience with his love of botany. In his letter of application he submitted his "Ultimate Aims for the Garden" and asked for five years to carry them out. With Gertrude Webster absent due to chronic illness, board member C.A. Dowdell presided over the selection process and deemed the "aims" too ambitious, instead hiring Marshall on only a two-month trial basis. Marshall became the third director of the Desert Botanical

W. Taylor Marshall became the Garden's third director in 1946. He revitalized the Garden after the war, and served until his death in 1957.

Garden on December 1, 1946, and it would take every bit of his expertise as a salesman to see his plans come to fruition.

Time for revival

As the Garden reopened in January 1947, Marshall's first task was to attend to the Garden's physical needs and the Society's finances.

Marshall needed someone who could do heavy labor and in early 1947 hired W. Hubert Earle as gardener. Earle, who suffered from severe asthma, had moved to Arizona from Indiana in search of a drier climate. According to his doctors, his only chance for survival was outdoor work and so he accepted the job. He and his wife, Lois, moved their trailer onto the Garden's grounds and made do with a very small salary.

The first post-war staff consisted of director W. Taylor Marshall (far right), his wife, Therese (far left), W. Hubert Earle (middle left), and his wife, Lois (middle right).

The lack of funds to pay Earle stemmed from the Garden's desperate financial situation. In truth, it was on the brink of ruin and there was serious talk about turning the Garden over to the state or the city. Instead, Gertrude Webster came to the rescue one last time. The Garden's long-time president and patron died on March 30, 1947, leaving her entire Arizona property to the Arizona Cactus and Native Flora Society for the continuation of the Desert Botanical Garden as a private institution.

Webster's will stipulated that in order to receive the bequest, the Garden had to have at least two hundred members. Upon hearing this, board member Lou Ella Archer and two other patrons promptly purchased seventy-five memberships each to give to friends. Their generosity made it possible for the Garden to qualify for Webster's endowment. It took more than two years for the estate to settle, but eventually $225,000 went into a trust fund for the Garden.

In the meantime, the Garden's board and small staff had to find ways to get by. To help make ends meet, the Garden's full-time hostess—Marshall's wife, Therese—set up "The Table" in Webster Auditorium. There she sold books, pamphlets, and curios to help fill the Garden's coffers while her warm, gracious manner helped win many friends.

Marshall set about reorganizing the Garden's governance and putting the Garden back in the public eye. Within six months of being hired, he had accomplished his first goal and was working on the second.

Marshall took an active part in making policy. It was his contention that re-

Lou Ella Archer was a founding member of the Garden and an ardent supporter. Archer House is named in her honor.

electing the Arizona Cactus and Native Flora Society's full Board of Directors each year was dangerous to the continued progress of the Garden. Under his direction, the bylaws were revised to provide board terms of four years, with one-fourth of the board being re-elected each year.

"This (was) to avoid the possibility of getting a board composed of all new members who would not be familiar with the work of the Garden," Marshall noted in the December 1950 issue of the *Saguaroland Bulletin*.

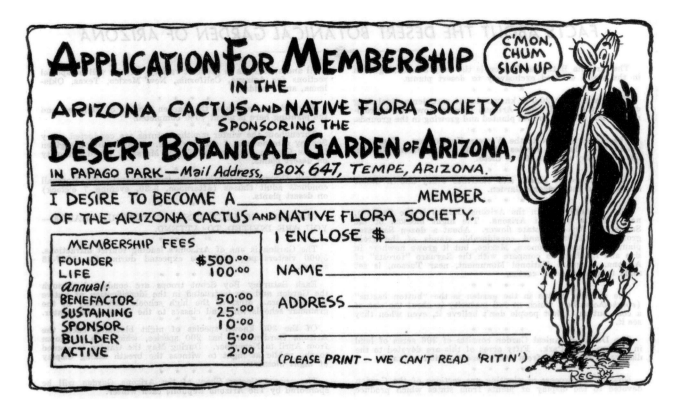

Students used the collection to learn about desert plants, circa 1950s.

Marshall also spearheaded the creation of an Advisory Board. The members of this board, according to the August 1947 issue of the *Saguaroland Bulletin*, were "specially qualified to advise the regular board in different matters and perform highly important work in focusing greater attention on Arizona, the desert, and the Desert Botanical Garden."

To put the Garden back into the public eye, Marshall approached Reg Manning, the nationally syndicated cartoonist for *The Arizona Republic*, who had been a Garden booster from the very beginning. Marshall and Manning enlisted the aid of *The Arizona Republic* and *The Phoenix Gazette* in supporting a promotion plan for the Garden. It included a membership

drive, an annual cactus show, and year-round publicity through stories and pictures in both papers. The papers agreed to handle and pay for much of the necessary printed material and help the Garden raise interest at the city, county, and state levels to improve the roads and signage into the Garden.

The Garden begins to grow

In June 1947 the Garden published its first bulletin, nine mimeographed pages without a name. The July issue bore the name *Saguaroland Bulletin*, submitted by Reg Manning in a contest to name the publication. Manning also designed a new look for the bulletin in 1950 and 1952.

The first cactus show on the Garden grounds took place in February 1948, ten years after the Society held its first show at Tropical Groves Nursery. Manning, who was chairman of the board, provided cartoon publicity, while *The Arizona Republic* began promoting the event seven months before it took place. The publicity was so effective that nearly eighteen thousand people attended the eight-day event.

Marshall led several collecting trips over the years with Hubert Earle, students, and other interested amateur and professional botanists. He continued the Garden's free public lectures and in 1948 instituted a series of Wednesday afternoon classes for adults.

Opposite/below: In 1952 superintendent Hubert Earle and his family moved into the new Archer House.

Opposite/above:. Reg Manning, cartoonist for The Arizona Republic, *contributed cartoons and designs for much of the Garden's early material. Manning was instrumental in creating awareness of the Garden in the larger community.*

This unique aluminum lath house was built in 1950 to house plants that needed protection from the cold. It was later devoted exclusively to cacti and became the Cactus House.

In 1950 Elwood R. "Jim" Blakley began assembling herbarium sheets for the Garden. Blakley had studied botany on the G.I. Bill at Arizona State College (ASC), now Arizona State University, and worked during weekends and breaks helping Marshall and Earle around the Garden. Having assembled a collection for one of his courses using specimens collected during trips with Marshall, Blakley had the necessary experience to start a herbarium. He became a full-time staff member upon graduation, first as junior botanist and then as curator of the herbarium.

Blakley worked on another major project in 1950, helping Earle build a long-awaited lath house. The structure was unique—an aluminum lath house built on a galvanized steel frame. It was funded by donations and would allow propagation of plants that required protection from the summer sun and cold winters. An anonymous $2,500 from "a board member" finally put the funding over the top and enabled the lath house to be built.

"Our ability to build it was a great thing," Blakley noted. "Many of Mr. Marshall's plants from his personal collection in Los Angeles were brought over to put in the lath house, and the display at the Garden was drastically increased."

Marshall and the board struggled with the ups and downs of membership throughout the postwar years. Charles Mieg, a local developer, aided their efforts. Mieg's interest and enthusiasm in reading about and collecting cacti led his wife, Lillian, to dub him a "cactomaniac," the title Mieg gave to a club he organized for the Garden in 1950. The Cactomaniacs met regularly for lectures, slide shows, potluck suppers, and field trips. Mieg, who was elected to the board in 1956, was generous with his time and funds and was the spark of the club for nearly a quarter century until his death in May 1974.

Desperately needed modern rest rooms were built on the rise east of the auditorium in 1951. In 1952 the Earle family moved into the Archer House, named for Lou Ella Archer, who had generously provided a great deal of funding and support for the Garden. The Earles and their two sons had lived in a small trailer, using the sanitary facilities in the auditorium, until the house was built.

Marshall spent ten-and-a-half years as director of the Desert Botanical Garden. Under Marshall's leadership, the Garden's living collection grew from a post-war 1,000 specimens to more than 18,000. Membership grew from nineteen to 408; six new structures were built; a herbarium was established; and programs for classes, lectures, and publications were begun. Marshall also donated his personal plant and library collections to the Garden.

On August 25, 1957, the Garden lost its devoted leader when W. Taylor Marshall died at the age of 71.

The Cactus House, the Garden's original aluminum lath house, helped protect vital parts of the living collection.

W. Hubert Earle
succeeded W. Taylor
Marshall as director
in 1957. Several new
buildings were
constructed under
his leadership.
Earle retired in 1976
after nearly thirty years
of service
to the Garden.

CHAPTER FOUR

CARE AND CULTIVATION

IKE MARSHALL, W. HUBERT Earle was self-taught in horticulture and botany. Earle's first exposure to cacti was at age four when in his native Winnipeg, Canada, he stepped bare-footed onto a cactus. His family moved to Indiana in 1923, where Earle completed his education and settled in the insurance business in Gary. In 1944 he developed such severe asthma that doctors gave him no more than three months to live unless he moved to a drier climate. Earle packed up his family, arriving in Arizona in 1945.

"Hubert Earle told me that when he was in Phoenix and he was out of work, he took his family through the Garden," Russell Haughey, who served as chief horticulturist under Earle in 1975-76, remembered. "He knew a lot of Latin and was explaining the Latin names of the plants to his family. Mr. Marshall came by and overheard this, started talking to Mr. Earle, and eventually offered him a job."

Earle went from gardener to chief horticulturist to superintendent. The board chose Earle as Marshall's successor, formally naming him director in May 1958.

One of Earle's first challenges was the effect of the sale of Papago Park by the state. Phoenix voters passed a bond election in 1957 allocating $1 million to develop a "class A" park and the city wanted the Papago Park area to do it. The state sold the park to Phoenix on February 25, 1959.

The original 1938 agreement leasing land to the Arizona Cactus and Native Flora Society was really no more than a permit, revocable by the state on one year's notice. In October the new owner took advantage of that by requesting—and getting—150 acres of the Garden's west section. In return, the State Land Commission offered a ten-year automatic renewal lease to the Garden for $25.

Between 1961 and 1963 the Garden gave up more land to the city in the Hole-in-the-Rock area to the south and a section to the north that was cut off by the construction of Galvin Parkway. The north section included Kriegbaum Knoll, where the founders did their first planting in

Opposite/above:
By the 1960s, the plantings around Webster Auditorium had matured considerably.

33

The Garden's gift shop did a brisk business in the Visitor Center, which was built in 1960.

1938. The trade-off was that the Desert Botanical Garden, left with a leasehold of approximately 145 acres, was made more accessible to the public and the improved quality of the surrounding park land made its position more secure.

Building a better Garden

Despite the income from the Webster trust and increasing membership, it had become necessary to sell the oriental rugs

and some of the antique furniture originally donated for Webster Auditorium to help meet expenses. "The Table" in the auditorium was not providing significant revenue. Before his death Marshall had talked of a separate shop near the entrance to help bring funds into the Garden.

Marshall's idea became reality under Earle's direction in November 1960 when the Visitor's Building was completed at a cost of $17,000. Access to the building and the Garden was made easier by the addition of a paved parking lot and black-topped path. The building was dedicated in February during the 1961 cactus show.

"We hear visitors comment on how easy it is to park, go through the building, visit the Garden, and return back through the building," the April 1961 *Saguaroland Bulletin* noted. "This new setup has caused our contributions and sales to go up 50 percent over the same period as last year."

Another innovation in 1961 was to keep the Garden open during the summer months. Until this time, the Desert

By 1999 the gift shop had added significantly more displays and merchandise.

A second aluminum lath house, this one for leaf succulents, was added to the Garden in 1965.

Botanical Garden was closed to the public during the summer, although it was always open to members and researchers. The break-even was set at getting two thousand people a month to come out in the heat; nearly double that actually visited the Garden that summer. It was deemed a success and the Garden was opened year-round.

Even with a new gift shop and summer hours, money was still tight, so Earle implemented yet another idea—voluntary admission fees. Adults would pay twenty-five cents and children ten cents, all on a voluntary basis. Members and their guests would be admitted free.

"A voluntary admission will not deprive those persons who wish to see the Garden, yet may be financially embarrassed," Earle noted in the June-July 1962 *Saguaroland Bulletin*. His plan was so successful that it added more than $9,000 in revenue to the Garden's $61,000 budget in its first year.

The Garden was able to add a second aluminum lath house in 1965 for succulents. The original lath house was then devoted exclusively to cacti, making it the "Cactus House" and the new one the "Leaf Succulent House."

The collection got a boost in 1966 when Shea Boulevard was extended. The eastern portion of the new road passed through property owned by the Page Land and Cattle Co. Fred Eldean, Sr., chairman of the company and a life member of the Garden, offered the right-of-way at no charge. The only stipulation was that the Desert Botanical Garden have the right to salvage the plants. This gave the Garden permission to collect legally protected plants from a strip of desert sixty feet wide and four miles long.

"John Weber and I and a couple of kids went in and gathered more than twenty-one hundred plants and brought them back to the Garden to plant them," said Lynn Trainum, who was a member of the

The Succulent House as it appeared in 1999.

Garden staff from 1965 to 1995. "We used rope, boards, and carpet to keep them from sticking us. It was quite a job."

Earle's staff was growing, too. John Weber became the Garden's horticulturist in 1958 when Earle was named director, and the two ran the Garden with the help of an assistant horticulturist, a couple of students, and a few volunteers. Once the Visitor's Building opened, William and June Hendrix were hired as the first maintenance and bookstore employees. A position for "interpretation" was created in 1964, which evolved into another book-

store and maintenance position in 1965. A secretary was added in 1966.

In true Garden tradition, volunteer help filled in some key areas. Armed with a New York editorial background, Lillian Diven took on the editorship of the *Saguaroland Bulletin* in 1964 and J. Harry Lehr, another self-taught botanist, became curator of the herbarium in 1970. Both were unpaid positions.

"It had to be done and I could do it, so I did it," said Diven, who also volunteered many hours in running the cactus shows. "I considered it my job."

Diven edited the bulletin as a volunteer until 1979. Lehr served as curator of the herbarium until 1984.

Significant donations

The Desert Botanical Garden's small library was greatly augmented when W. Taylor Marshall added his own collection to it. Marshall had thought his to be the finest collection of its kind anywhere until he came across Max C. Richter, owner of the Book Den in Santa Barbara, California.

Richter had, over forty years, assembled one of the most complete collections in the world of cactus and succulent books and botanical prints. Marshall cultivated Richter's friendship, a relationship that Earle continued after Marshall's death. Richter talked about donating his collection to the University of California in 1959, but Earle persuaded him that it would be far more appreciated and use-

ful at a small, specialized institution. In 1965 Richter resisted an eleventh-hour offer from UC Berkeley of $78,000 and began an annual donation of material to the Desert Botanical Garden.

As part of the agreement, Earle promised to build a library named for Richter at the Garden. Designed by Les Mahoney, a long-time member who also designed the lath house, Archer House, and the Visitor's Building, the library was completed in February 1968 at a cost of $22,500. It took more than a year to furnish the library and fill the shelves, but at last it was dedicated on April 26, 1970. Richter, whose ill health kept him from attending, expressed his gratitude in a letter to the board, noting "the able manage-

The Eliot Patio was named in honor of William Eliot, who served as the Garden's attorney in the 1950s. Eliot died in 1959. The "Cactomaniacs" planted the bed pictured in the foreground.

37

Desert Botanical Garden

Above left: The flowering agave became the Garden's new emblem in 1976.

ment of the former W. Taylor Marshall and the no less able W. Hubert Earle." Richter died on April 14, 1973.

In January 1968 the Garden lost its chief volunteer when Earle's wife, Lois, died suddenly. For many years she had helped support the family by working at a downtown department store, then in the president's office at Arizona State University. In her "spare" time she served as the Garden's bookkeeper, storekeeper, and hostess.

Lois Earle's death left the Garden with no one to take care of the many details she handled. Member Fred Beselt asked Hubert Earle to make a list, then Beselt sent out an appeal for help—typing, filing, cleaning seed, making plant labels, and other tasks. About thirty people responded, creating a loosely organized "Service Group" to help fill the gap.

Many of the Garden's members and friends also contributed money for a memorial to Lois Earle. After many months of discussion, the board decided to use the funds to acquire twenty herbarium cases for its new herbarium building, which was completed in January 1972. Named in honor of the Garden's hardworking hostess, the facility was dedicated as the Lois Porter Earle Herbarium on April 16, 1972.

In April 1968 the Garden lost another staunch supporter when Lou Ella Archer died. A long-standing member who was board president at the time of her death, Archer's contributions would be missed.

More growth and changes

As the Garden's facilities grew over the years, so did the number of members. From the meager nineteen who lent their support to the Garden's post-war reopening in 1946, the list grew to more than one thousand by the beginning of 1970.

"Increase in membership has been due to many factors: enthusiasm of existing members, newspaper and magazine articles, lectures here and elsewhere, the classes given here at the Garden, and of course the visitors to the Garden," Earle noted in the February 1970 *Saguaroland Bulletin*. "Remarkable, when you realize that we have never had any organized membership campaigns."

Another "first" occurred in 1971 when Dr. Howard Scott Gentry, a noted plant explorer, joined the Garden staff as research botanist. His primary support was a three-year National Science Foundation grant the Garden received in 1972 for Gentry to conduct research on the genus agave. He published his definitive *Agaves of Continental North America* in 1982. Gentry also received a small stipend from the Garden and rent-free living quarters in Archer House, which was vacated by Hubert Earle when he remarried in 1970 and moved to Phoenix.

The Garden lost yet another friend in 1976 when John Eversole, chairman emeritus of the Garden's Board of Directors, died. He had joined the board in 1947 and served as chairman from 1950 to 1975.

Lois Earle (left), wife of director Hubert Earle, won a host of friends for the Garden with her warm personality. The Lois Porter Earle Herbarium was named in her honor.

Eversole, an attorney, represented the Garden in the settlement of Gertrude Webster's estate, devoting many hours of legal service at no cost.

The flowering agave became the Garden's new emblem in 1976. Created by the Roberts Group, it was part of a move to develop a corporate image for the Garden.

"Like other institutions, the Garden realized the advantages of a readily identifiable symbol. We needed something appropriate that would be attractive to our members, our friends in the botanical world, our visitors, and the public at large," the August-September 1976 *Saguaroland Bulletin* explained. "We think our agave fills the bill. It is a succulent, it is graphic,

and no other organization uses the agave as a symbol."

A year of endings and beginnings, 1976 marked the end of an era when W. Hubert Earle retired on October 1. Earle, who in 1945 was told he might have only months to live, had given the Garden nearly thirty years of dedicated service. Earle passed away on December 12, 1984.

"He was very thoughtful and caring about his employees and the Garden," long-time staff member Lynn Trainum recalled. "He used to go out at night and give lectures and speeches and if they paid him, he'd turn the money over to the Garden. He was a very nice guy."

Cholla and organ pipe cactus frame the Visitor's Building at the Desert Botanical Garden, pictured here in 1974.

CHAPTER FIVE

BRANCHING OUT

STARTING WITH JOHN WEBER in 1958, Earle had brought in help as funds allowed. By the end of 1975 the staff numbered fourteen which, augmented by eighteen volunteers, fulfilled the Garden's mission and served a membership of nearly fourteen hundred.

One of those staff members was Rodney Engard, who went to work at the Garden as a twenty-two-year-old student horticulturist in 1971. Weber recognized a kindred spirit in Engard and began training the promising student to take over Weber's own job. When Weber left the Garden in 1973 for a position at The Phoenix Zoo, Engard was named chief horticulturist. He became superintendent in 1974.

Earle's retirement left the Board of Directors with a dilemma. As superintendent, Engard was the likely successor, but only Mildred May, president of the board, believed wholeheartedly in Engard's ability to handle the job. His

Rodney Engard, who succeeded Hubert Earle as director in 1976, brought an energetic new vision to the Garden.

youth made the other board members cautious, so he was offered the position of acting director. Engard took the job but refused the title. He directed the Garden's affairs as "superintendent" for nearly two years before he was given the formal title of director in 1978.

Above: The 1977 Cactus Show, displayed in Webster Auditorium, drew twenty-five thousand visitors to the Garden.

Below: The Garden's docents provided visitors with a close-up, in-depth look at the desert during daily tours.

A fresh vision

One of Engard's first moves was to declare the Garden's support for the Arizona Native Plant Law. As part of its support of the law, the Garden would no longer sell plants removed from native habitats; instead it would only sell those that were nursery-propagated. This led to the creation of a propagation department in 1977 and a propagation greenhouse in 1978.

Engard and the board instituted a number of policy changes, many of them to help improve the Garden's finances. One was to begin charging admission to non-members in 1977, despite fears that it would discourage attendance. Those fears proved to be groundless as very few balked at paying a dollar for adults and 50 cents for children. The Garden also began charging fees for classes, holding special plant sales, and actively seeking corporate memberships.

Many of the Garden's rare botanical prints were mounted in an exhibition in September 1977 as part of the fund-raising effort. Lillian Diven, the volunteer editor of the *Saguaroland Bulletin*, took on the task of identifying the prints. She spent hours poring over major reference works on botanical illustration, resulting in a complete documentation of the collection.

Volunteers such as Diven played an increasingly important role in the energetic Engard's plans. Carolyn Hoppin took on the task of volunteer public relations director, effectively getting attention for the Garden from the media and the public. In November 1977 Engard and Hoppin crafted plans for an "AridZona" contest, sponsored in conjunction with Salt River Project, to recognize landscap-

ing using drought-resistant plants. The contest was such a success that it became an annual event.

Another area Engard tackled was education, naming horticulture staff member Sherry Couch Krummen as director of the new education department in 1976. Krummen wrote pamphlets on desert gardening and, with chief horticulturist Russell Haughey, planned and conducted Saturday field trips. Perhaps the new department's most important accomplishment, however, was the creation of a docent program for the Garden.

Until Engard became director, there was little done in the way of guided tours of the Garden. In the early days the director or an available member would take groups through, but there was no cohesive program of tours or trained people to conduct them. Krummen tried to handle requests for tours herself but it was too much for one person, so she began organizing the docent program.

Twenty-seven people enrolled in the first docent training program, which began in October 1977. After five classes, an oral and written exam, and a field trip, the new docents received smocks and aprons sporting the Garden's logo and began leading tours. From that beginning, the docent program grew to become a key part of the Garden's public interpretive program. It also became the basis of the Garden's organized volunteer program, setting the model for an orderly effort to recruit and train volunteers to work in all areas of the Garden.

The Desert Botanical Garden's holiday tradition of *Las Noches de las Luminarias* began as Luminaria Night in December 1978. The Garden's "holiday gift to the public," it was a one-night event involving a thousand paper bags and candles and many hours of staff and volunteer time. A mariachi band played in Eliot Patio and hot cocoa (replaced by hot cider the next year) and cookies, baked by staff and volunteers, were served in the auditorium. So many people attended that they created a gridlock on Galvin Parkway.

In later years the Garden began charging admission for Luminaria Night. By 1998 the event was spread over four days and required seventy-four hundred luminarias, more than twenty-nine thousand candles, five-and-a-half tons of sand, 2,667 dozen cookies, and more than a thousand gallons of cider.

Changes in governance

In 1979, under Engard's leadership, several changes were made to the Garden's governing structure. The first was a new corporate name. No longer would the Garden be sponsored by the Arizona Cactus and Native Flora Society; instead, it became Desert Botanical Garden, Inc., a non-profit and non-tax-supported institution. The new entity would be the official lease-holder on the land from the city.

Mooch was the director of rodent control in 1978. Dr. Gentry's cat, Buster, also was among the Garden's beloved feline sentries.

Dr. Charles Huckins, director from 1979 to 1983, sought and received accreditation for the Garden from the American Association of Museums.

Until this time, the Garden was governed by an Executive Board and an Advisory Board. The bylaws were amended in 1979 to combine the two boards into a single Board of Trustees. Previously, there had been no limit to how long a member could serve on a board; now members would be limited to two consecutive three-year terms. They could be re-elected after a one-year absence from the board.

The board also established a finance committee to help stabilize the Garden's funding challenges and a development committee to work with the staff on a master plan.

Rodney Engard had taken on the directorship of the Garden with a clear vision for the Garden's growth and development. By 1979, with the help of dedicated staff, volunteers, and board members, he had set many of his plans into motion. Engard's first love, however, was botanical exploration. He left the Desert Botanical Garden in September 1979 to pursue advanced studies in botany, and later became the first director of the Tucson Botanical Garden. Engard died in April 1990 at the age of 42.

"Rodney Engard had tremendous foresight," noted Wendy Hodgson, who joined the staff in 1974. "He had an incredible mind and was a very good botanist. He also was ahead of his time."

Luminaria Night began in 1978 as the Garden's holiday gift to the public.

Staff member Lynn Trainum manned the gift shop in the Visitor's Building, which also served as the entrance to the Garden when it was built in 1960. It continued to be the gateway to the garden until 1990.

New directions

Engard's successor was Dr. Charles Huckins, the Garden's sixth director and the first to hold a doctorate in botany. Huckins came to Arizona from the Missouri Botanical Garden, where he chaired its Indoor Horticulture Department.

The new director immediately outlined his agenda for the Garden: the development of a master, long-range plan; expansion of the volunteer program; and creation of a broader base of financial support.

The latter item required a shift in attitude at the Garden.

"When I arrived, the people associated with the Garden took pride in not being supported by tax dollars," Huckins recalled. "Being a rather pragmatic person and appreciating what the potential could be from many directions of support, we took a different approach. We went after corporate support and government grants."

Huckins instituted a more aggressive policy toward increasing revenue. Firmer controls were placed on expenses while fees for membership, admission, and classes were increased. Sales in the gift shop improved with the introduction of higher quality merchandise, and plant sales were better publicized and attended. The Garden also took more assertive action in increasing personal memberships.

Huckins obtained a grant from the Institute of Museum Services (IMS) in September 1980, the first time the Garden had sought funds at the federal level for operations. The $35,000 grant went toward educational programs, better protection for the library and herbarium collections, and getting plant records in order. IMS grants in subsequent years enabled the Garden to computerize its plant records and supported other operating needs.

Volunteer recruitment and activity expanded during Huckins' tenure. He added a full-time staff person to coordinate volunteer activities and increased recognition programs for volunteer efforts.

Visitors inspect the entries in the 1983 cactus show.

Strengthening the foundations

In 1982, following preliminary work by Huckins and the board's development committee, the Garden contracted with consultants to produce a comprehensive long-range plan. Over the next few years it would be discussed, revised, and reworked as it was implemented.

"Even though we had professional assistance in putting it together, it was really work done by the staff, board, and volunteers at the Garden," said Huckins, who left before the plan was implemented. "Victor Gass, who was a key member on my staff, told me some years after I left that I would be pleased because most everything we set out to accomplish in the plan had been accomplished."

The Garden was accredited by the American Association of Museums (AAM) in April 1983. The accreditation process included a detailed written questionnaire and a site visit by members of the AAM Accreditation Commission. The Desert Botanical Garden was one of only twelve botanical gardens to receive the designation.

Huckins noted that accreditation by the AAM gave the Garden the opportunity to demonstrate to the community "the importance of collecting, conserving, displaying, and interpreting precious living resources which are every bit as important to man's cultural enrichment as the finest objects of art history."

That recognition helped achieve another long-term goal in November 1983 when the Desert Botanical Garden was designated a beneficiary of the Combined Metropolitan Phoenix Arts and Sciences (COMPAS). COMPAS raised funds and attention for the Heard Museum, the Phoenix Art Museum, the Phoenix Symphony, and the Phoenix Zoo.

"It had been a long, concerted effort over several years," Huckins said. "Henry Triesler, who was on the board that recruited me for the director's job, also was involved in the Heard and local art museums. He felt that the Garden ought to work toward becoming a beneficiary of

COMPAS, so I give him a lot of credit for encouraging us to pursue it."

Huckins resigned at the end of 1983 to become director of the American Horticultural Society. Under his leadership, the Garden had increased its stature as a museum and scientific institution and set long-term direction for its future.

Nancy Swanson, who joined the Board of Trustees during Huckins' tenure, appreciated the leadership and dedication he brought to the job.

"I was brand new to the board with others who had been there forever. He was wonderful in helping me get started. He had so much to contribute and had the Desert Botanical Garden very much in his forethought."

Changes at the Garden

Frederick W. Shirley, a retired U.S. Army lieutenant colonel, succeeded Huckins in January 1984. Shirley, formerly public affairs officer at Fort Knox, Kentucky, held a master of business administration degree from ASU.

As Shirley joined the staff, a long-time member was leaving it. J. Harry Lehr, who had served as curator of the herbarium since 1970, retired after having built a nationally recognized collection. Wendy Hodgson, who had served as illustrator, education assistant, and herbarium assistant for ten years, was appointed to succeed Lehr.

The Garden created a memorial to a long-time member in October 1984 when the Rhuart Demonstration Garden, an area designed to demonstrate home desert landscaping, opened. It was dedicated to John Rhuart, who had joined the Arizona Cactus and Native Flora Society in 1935

Frederick Shirley, a retired Army officer, was director of the Garden in 1984-1985.

and remained a staunch supporter of the Garden until his death in 1982.

Rhuart served as president of the board from 1970 to 1976 and then as chairman until 1981. In 1978 he donated funds for thirteen hundred feet of pipeline west of Archer House. The line made it possible to test plant species in that area, which later became the site of the garden dedicated to him.

Under Fred Shirley, Archer House was renovated, the Garden's strategic plan was activated, an earlier attempt to publish a Garden magazine was successfully revived, and the gift shop was overhauled. In the fall of 1985 Shirley left to pursue other interests. Once again it was time for a change.

This crested saguaro is the most photographed plant at the Garden. Crested saguaros are a mutant form of the cactus and are rare.

The Collections

Color photos by Jennifer Johnston

Woolly-headed Barrel Cactus
(Echinocactus polycephalus)

Santa Rita Prickly-pear Cactus
(Opuntia santa-rita)

Prickly-pear Cactus
(Opuntia a . basilaris)

Golden Torch Cactus
(Echinopsis spachiana)

Facing Page:
Lady Finger Hedgehog Cactus
(Echinocereus pentalophus)

THE SUCCULENT COLLECTION

Right:
Aloe rubroviolacea

Far Right:
**Golden-flowered
Agave**
(*Agave chrysantha*)

Above: **Parry's Agave** (*Agavi parryi*)

Right: **Aloe ferox**

Facing Page: **Parry's Agave** (*Agave parryi* **var.** *truncata*)

DESERT WILDFLOWERS

Above: **California Poppies**
(Eschscholzia californica)

Right: **Bluebells and California Poppies**
(Phacelia campanularia, Eschscholzia californica)

Left: **Penstemon and California Poppies**
(Penstemon parryi, Eschscholzia californica)

Above: **Bluebells (Phacelia campanularia)**

Facing Page: **Desert Primrose** *(Oenothera caespitosa)*

CHAPTER SIX

A THRIVING LANDSCAPE

I N 1985 Dr. Robert Breunig was deputy director and chief curator at the Heard Museum in Phoenix. He also was just the person the Garden's Board of Trustees was looking for.

"We realized that we needed a scientist, and we needed to have someone who could articulate a vision," said Nancy Swanson, who served as board president from 1984–1985. "We approached Robert Breunig at just the right time and place. He wanted to be a director and we wanted a scientist who had experience in a museum."

Breunig, who held a doctorate in cultural anthropology, came to the Garden with a clear picture of what he wanted to accomplish.

"One of my goals was to make the Garden a vital institution," Breunig said. "I wanted people to see it as a fundamental resource for living and working in the desert. If the Garden was dynamic and effective, then it would help with the larger perception of the desert as an ecosystem."

Just prior to Breunig's arrival, the Garden joined the Center for Plant Conservation, which had been formed by a group of botanical gardens to identify and conserve endangered plants. The affiliation with the center was a good fit with the new director's vision, and the Garden would later become one of the country's leaders in terms of plant conservation.

Breunig believed that research should be a critical component of the Garden and that the Garden would have little credibility unless it helped to generate more knowledge about the desert ecosystem. His one reservation about taking the job was his own credibility, since he was coming from a non-botanical field and was not himself a botanist.

"I didn't know if I had the credibility to run an important botanical institution, so I made a deal with the board that if they hired me, they would allow me to hire a first-rate assistant director who would

Dr. Robert Breunig, director of the Garden from 1985 to 1994, led an effort to improve the research, horticulture, and education programs at the Garden.

have that credibility," Breunig said. "They agreed and the first day on the job I called Gary Nabhan, who came the following spring. He really put our research program on the map."

Dr. Gary Nabhan, who was widely recognized for his studies of desert plants, came to the Desert Botanical Garden in February 1986 from the University of Arizona Office of Arid Land Studies. Over the next six years, as director of research

for the Garden, he would be recognized nationally for his research and conservation work.

Nabhan's arrival at the Garden coincided with the passing of a long-time supporter of the Garden. Reg Manning died in February 1986 at the age of 81. The nationally known cartoonist had helped promote the Garden and educate the public through his work in *The Arizona Republic* by poking good-natured fun at the average person's introduction to prickly plants. Manning was chairman of the Garden's Executive Board from 1947-1952 and remained a loyal friend of the Garden until his death.

Making things happen

Along with raising the level of the research effort, Breunig focused on improving the horticulture program. Under his direction, documentation of the living collection was improved. He and his staff devised a system to better plan and manage the living collection over the long term, instituting an annual collection assessment and processes to fill the "gaps."

Another of Breunig's aims was to strengthen the quality of the education program.

"We had a dynamic director of education, Kathleen Socolofsky," Breunig noted. "We asked ourselves what the educational content of the Garden should be, what people should learn when they came here, how they should learn it, and how the message should be organized."

Out of those discussions came the idea for an interpretive master plan for the Garden. The plan was aimed at making sense out of the Garden's physical layout,

Dr. Robert Breunig (left) worked with fellow staff members and volunteers to get the Garden ready for the Luminaria event in 1986.

Garden horticulturists plant a crested saguaro, which was placed in the Garden in 1992 after authorities confiscated it as evidence in a cactus-rustling case.

eventually producing the idea for the Garden's looped trail system.

Ruth Greenhouse, a research associate who later became part of the education department staff, was very involved in the planning. Greenhouse also had proposed an idea several years earlier for an ethnobotanical exhibit, which had been included in the long-range plans by Charles Huckins, but there had been little money to develop it. Under Nancy Swanson's leadership in 1985, the Board of Trustees designated the project as its number one capital goal and began to raise funds. Upon his arrival at the Garden, Breunig suggested applying to the National Endowment for the Humanities (NEH) for a "big grant."

"It takes the right people to make things happen," Greenhouse said. "It was neat synchronicity and timing. When Robert came on, he said 'Okay, you're going to make this happen.' He was great at getting things done, really good to work with, and a great delegator."

Greenhouse's project was the Plants and People of the Sonoran Desert Trail, an exhibit that was unique in its time. Completed in 1988, it became a model for ethnobotanical exhibits at many other institutions. It made enough of an impact that the American Association of Museums asked Breunig to testify before Congress on behalf of the NEH about the success of the exhibit.

As the Garden approached its fiftieth anniversary, one long-standing tradition saw its last year. The annual cactus show, a fixture at the Garden every spring since 1947, had its last appearance under the Garden's sponsorship in 1987. The event had become too much of a drain on the Garden's resources and manpower, and the Garden withdrew from active participation. The show had been held for a time in conjunction with the Central Arizona Cactus and Succulent Society (CACSS), which was formed in 1973. The CACSS continued to host the show at the Garden on its own after 1987.

George Lindsay, the Desert Botanical Garden's first director, returned to the Garden to share memories with trustees, staff, and volunteers at a dinner in January 1987 marking the fiftieth anniversary of the Arizona Cactus and Native Flora Society's incorporation. The Garden celebrated the fiftieth anniversary of its dedication in February 1989 with a two-day event that featured activities for adults and children.

In 1990 the Garden published the final edition of the *Saguaroland Bulletin* and discontinued the quarterly *Agave* magazine. *The Sonoran Quarterly*, a new bulletin edited by volunteer Carol Schatt, replaced the older publications in the spring of 1991.

Restoring and improving the facilities

When Breunig arrived at the Garden in 1985, he found a physical facility in need of attention. Overhead power lines marred the landscape, power and water supplies were inadequate, the trails were beginning to break down, and some of the buildings were in poor condition. In

The looped trail system allowed visitors to find their way easily through the Gardens's many paths and exhibits.

Dr. George Lindsay (left), the Garden's first director, and Dr. Robert Breunig cut the cake at a celebration in January 1987 marking the fiftieth anniversary of the incorporation of the Arizona Cactus and Native Flora Society.

1988 Phoenix voters approved a bond package that provided $1 million for improvements to the Garden's infrastructure, which is on city-owned land. It included a new water line, new parking facilities, underground power lines, and the restoration of Webster Auditorium. Galvin Parkway was widened during the same time period, relocating the Garden entrance and creating a safer traffic situation.

The Webster Auditorium restoration included a new roof; new heating, cooling, and plumbing systems; repair of damaged beams and ceilings; and many other improvements. A catering and concession kitchen and the Ullman Terrace, a three-level patio, were added to its

south side, replacing a carport and a crumbling parking lot. Ullman Terrace, which included a cafe, was named in honor of trustee Virginia Ullman, who challenged the Board of Trustees to match her donation. The restoration was completed in 1990 and the building was entered into the National Register of Historic Places.

The Fleischer Propagation Center also was completed in 1990. Funded by a grant from Morton and Donna Fleischer, the facility included two greenhouses and a propagation building. It enabled the Garden to upgrade the management of its

Members of the Gila River Pima Indian Community renovate an authentic roundhouse as part of the Plants and People of the Sonoran Desert Trail.

plant collection through propagation of new plants and rehabilitation of ailing ones.

The Desert House, a 1,657-square-foot house built on the Garden's grounds, was created as the key element in the Center for Desert Living in 1993. A cooperative project by the City of Phoenix Water Department Conservation Office, Salt River Project, University of Arizona Office of Arid Land Studies, Bank One, Arizona Department of Commerce Energy Office, and the Desert Botanical Garden, the Desert House was conceived as a research project on conservation. The study would determine how people could

live comfortably and economically in hot, arid regions with all the normal amenities while conserving natural resources.

Various families were selected to live in the Desert House, each for an extended period of time. Their water and power use was monitored and recorded to discover if the house's design and materials, not the behavior of the occupants, would contribute to saving water and energy. By 1999 the results from three families revealed that the house's design did save as much as 40 percent when compared to similarly sized conventional homes.

"We wanted it to be a real-time test of desert living, so the house is full of meters, sensors, and all kinds of stuff,"

Breunig said. "The family in residence tries to live as normally as possible with two hundred thousand people walking around outside."

A nearby information center allowed visitors, via interactive video, to see how the house was designed and to take a "tour" of its interior.

In 1992 a $634,000 National Science Foundation grant provided what Breunig described as "the keystone funding in the expansion and upgrade of all the Garden's interpretive services." It helped provide a general upgrade of the Garden's path system, signage, "investigation stations," a new exhibit orientation area, and materials for use by local schools. The Garden's loop trail was reconstructed as the Desert Discovery Trail as part of the five-year project.

When Robert Breunig arrived at the Desert Botanical Garden in 1985, he was set on redefining the image of the Garden

in the larger community and his vision took the Garden to a new level of professionalism and stature. After nine years at the helm, however, he was ready for a new challenge. Breunig accepted the position of executive director for the Santa Barbara Museum of Natural History in California in the fall of 1994.

"It's fair to say that we took it to another level, with the emphasis on 'we,'" Breunig noted. "I was blessed with a first-rate staff. They were, and continue to be, a great group of people—every one of them talented in their own right."

The September-November 1994 *Sonoran Quarterly* observed that Breunig "led us to be more vigilant of our collection, more interested in our visitors, more concerned with each other, and more secure in how successful we could become."

Ullman Terrace, named in honor of trustee Virginia Ullman, was added to Webster Auditorium during a restoration project in 1990.

Carolyn Polson O'Malley succeeded Robert Breunig as director in 1995. Under her leadership, the Garden drew up a master plan to guide its first-ever capital campaign in 1999.

Carolyn Polson O'Malley signs on

As part of the effort to increase the Garden's visibility, Breunig had hired Carolyn Polson O'Malley in November 1993 as assistant director. O'Malley's background was in public relations and non-profit management and she was hired with that focus in mind. When Breunig left the Garden, the Board of Trustees wanted to take their time in finding a worthy successor. O'Malley was named acting director in the interim.

O'Malley was officially named executive director in August 1995. In addition to overseeing the final stages of major projects and improvements initiated under Breunig, O'Malley began to set an agenda of her own.

"Like Robert, we focused on benefits. He concentrated on benefits to the visitors, giving us greatly improved visitor amenities," O'Malley noted. "I also began looking at rewards to the employees in terms of benefits and facilities."

A major grant from the National Science Foundation in 1992 helped fund research that made the Garden's signage more interactive and effective.

The first major task O'Malley and the board tackled was a long-range master plan that took three years to develop. In the process they developed a vision statement and identified two goals to help the Garden reach that vision—the need to increase the Desert Botanical Garden's visibility in the community and the need to increase revenues.

In March 1997 the Desert Botanical Garden celebrated the completion of its five-year, $2 million series of renovations that enhanced the beauty of the Garden, increased comforts for visitors, and enriched understanding of desert plants. The results included a new hierarchy of trails with hardscape improvements to the Desert Discovery Trail, redesign and renovation of the Sonoran Desert Trail, and new trailheads for the Plants and People of the Sonoran Desert Trail. There also were new visitor wayfinding signs, new shade islands and chilled drinking fountains, and new interactive exhibits, trail guides, and educational materials.

The Garden launched a Desert Landscaper Certification Program in 1997. The program offered "hands-on" learning in all aspects of desert landscape installation and maintenance taught by experienced Garden horticulturists.

The Garden unveiled a new website in 1998 with the support of Bull Information

The Desert House was built in 1993 to study how people could live comfortably and economically in hot, arid regions with all the normal amenities while conserving natural resources.

Systems. Begun as an "electronic brochure," it provided information about the Garden and its services. Eventually the website would offer access to information about the plants in the collection.

Once the long-range planning process was completed, the Garden launched its first-ever major capital campaign in January 1999. The goal of the "Growing a Legacy for Generations" campaign was to raise $15.7 million by the end of the year.

That figure was a far cry from the few thousand dollars the Arizona Cactus and Native Flora Society had scraped together to found the Desert Botanical Garden. That the Garden could conceive and execute a major capital campaign of that magnitude was both a measure of how the Garden grew and prospered in its first sixty years and a tribute to all those who guided the Garden to success.

Over the years the Garden's leaders built a foundation of financial strength and stability while increasing the staff, improving the infrastructure, and adding trails, exhibits, and structures. By 1999 the Garden had become a scientific institution that took a very business-like approach to achieving its goals. The annual operating budget was $4,150,658, with revenues coming equally from three sources: admissions, retail sales, and contributions (membership, fundraising, and grants). Membership numbered more

This thatched ramada provided a shady retreat on the Plants and People of the Sonoran Desert Trail (photo by Steve Priebe).

By 1999 retail sales from the gift shop, along with plant sales, contributed one-third of the Garden's revenue.

than nine thousand and yearly attendance exceeded 250,000 visitors. The staff had grown to sixty-four full-time and seven part-time people, while 525 volunteers contributed approximately 60,000 hours.

As the Desert Botanical Garden prepared to celebrate its sixtieth anniversary in 1999, O'Malley observed that the Garden has remained true to its original mission.

"Gertrude Divine Webster would be really smiling now if she knew that we are serving the very same mission that she outlined sixty years ago."

This magnificent display of agaves is part of the site where Ullman Terrace was built in 1990.

CHAPTER SEVEN

THE COLLECTIONS

THE ARTICLES OF INCORPO-RATION for the Arizona Cactus and Native Flora Society outlined the mission for the Desert Botanical Garden in 1937. The Garden was "to exhibit, to conserve, to study, and to disseminate knowledge" about the arid-land plants of the world. Over the years, the Garden's collections conformed to and supported that mission despite financial challenges and leadership changes.

The Living Collection

The Desert Botanical Garden's first plant accession, a creeping devil cactus (*Stenocereus eruca*), was planted in 1939 in bed number fourteen just outside Webster Auditorium where it was still growing strong sixty years later. It was one of the five thousand cacti and other desert plants introduced into the Garden within the first months of its founding.

George Lindsay, the Garden's first director, brought back more than four hundred plants from a trip to Baja California in August 1939. Among the treasures he introduced were boojums (*Fouquieria columnaris*), cardons (*Pachycereus pringlei*), and copalquin (*Pachycormus discolor*), which made an awe-inspiring forest of Baja natives near the northwest corner of the auditorium.

Over the years specimens were added to the Garden from hundreds of resources through collecting trips, exchanges with other institutions, gifts from members and scientists, and salvage operations such as the Shea Boulevard expansion. The addition of lath houses and greenhouses allowed a wider variety of plants to be propagated and displayed. The collection grew into an extremely significant one, according to Robert Breunig.

"Its strengths are the agaves and, of course, the cacti," he said. "I think the agave collection is more complete, although the cacti collection is considered one of the best public collections in the world."

In addition to protecting plants, the Cactus House, pictured in 1999,
offered an interesting and informative exhibit for visitors.

As the Garden's collection grew, so did the problems of collecting and managing the collection. Collecting specimens became more difficult as the United States and Mexico placed more restrictions on the removal of native plants. Harsh growing conditions, disease, and frequent rabbit attacks sometimes challenged the plants' survival. But from the first, close attention has been paid to documenting the collection.

"From a scientific point of view, the value of a collection rests in the data that goes with it," Breunig noted. "George Lindsay put the Garden on the right path. He understood the importance of keeping good data and established standards for collecting data while collecting plants, which continue to this day."

Initially there was no set system for managing the collection. It varied from administration to administration. Serious organization of the collection began in the late 1960s under Hubert Earle when planting beds were assigned numbers. Rodney Engard prepared the first map of the collection and Sherry Couch Krummen developed records on the plants.

One of the IMS grants secured by Charles Huckins enabled Victor Gass, curator of the living collection in the 1980s, to switch the Garden's plant records from a ledger book/card file system to a computer. The computer made capturing, organizing, and using the information much easier and more thorough, and immeasurably increased the ability to analyze data.

Dr. Liz Slauson, who became curator in 1990, instituted an annual inventory to see what was alive, what had died, and what was needed for representation in the future. Slauson was instrumental in designing a program for conserving and managing the collection over the long term through recollecting, propagation, and conservation.

The many restrictions placed on collecting and the rarity of some plants meant that the Garden must rely more and more on propagation. The addition of greenhouses, propagation facilities, and a seed vault in the 1990s was vital to the Garden's ability to maintain the diversity of its living collection.

Throughout the Garden's history, its horticulture staff played a vital role in maintaining the collection. Charged with tending the plants on a daily basis, horticulture staff members and volunteers also developed plant exhibits, designed and redesigned beds, planted and transplant-

ed specimens, helped in salvage projects, and experimented with growing techniques. They often were involved in building trails and facilities as well.

The Center for Plant Conservation

The Center for Plant Conservation (CPC), a national consortium of botanical gardens and arboreta, was organized in 1985 as a response to the rapidly increasing losses of plant species. Its mission was to identify, research, and rescue plant species that had become rare or were in danger of extinction.

The Desert Botanical Garden, a founding member of the CPC, joined in the effort to collect germplasm of rare and

A moon cactus (Harrisia Martinii) *added beauty to the Garden with its display of fragrant blossoms.*

This fishhook barrel cactus was among the plants salvaged from the Shea Boulevard expansion in 1966.

endangered plants to serve as a backup against extinction in the wild. The Garden also studied many of these species in the wild to better understand the threats to the species, how they might be better conserved, and how they might be grown and propagated. This special collection of plants is an important component of the Garden's mission to conserve plants.

The Max C. Richter Memorial Library

From the beginning, researchers at the Desert Botanical Garden relied on books and other printed matter to help them pursue their studies. The Garden's original book collection, which consisted of about eighty books and papers provided by some of the founders and housed in the director's office, did not offer a great deal of help.

Director W. Taylor Marshall greatly augmented the collection in the 1950s with the donation of his five hundred volumes on cacti and succulent plants. Max C. Richter's decision to donate his extensive collection to the Garden in 1965 provided an even more significant advance. In addition to his own books, Richter had acquired the collections of Scott Haselton, editor and publisher of the *Cactus and Succulent Journal of America*, and of P. Montandon, editor of *Cactus*, and these volumes also became part of the Garden's library. The dedication of the Max C. Richter Library in 1970 further boosted

the value of the collection, offering researchers a proper facility in which to use the Garden's materials.

Funding was slim for the new library. Eleanor Radke, Peg Gill, and Elizabeth Fritz served in succession as volunteer librarians before Rebecca Henry was hired as the first part-time paid librarian in 1979. Many volunteer hours were spent in cataloguing the Richter books along with a collection of more than five hundred botanical prints.

The Richter Library functioned as the botanical information retrieval center for the Garden. As the collection grew, so did the demand for its use. Reciprocal agreements with other libraries, museums, and horticultural organizations greatly increased the reach and usefulness of the Desert Botanical Garden's library collection. The advent of the computer and on-line access further enhanced its stature.

In 1995 the library received another priceless gift when the family of Lyman Benson, the renowned cacti systematist, donated his collection of correspondence, manuscripts, and books to the Garden. Dr. Edward "Ted" Anderson, a Garden staff member, had been a student and friend of Benson and was instrumental in securing the Benson collection.

By this time, the library was running short on space. Jane Cole, who served as librarian from 1983 to 1997, received an Institute of Museum Services (IMS) grant in 1995 to install compact shelving. This increased the storage space and helped better conserve the collection. Dianne Bean, who succeeded Cole, began an effort to preserve older material by scanning it into a computer. Under her direction, a computer database was built to make the library's thousands of titles easier to search.

A unique aspect of the library is its rare book collection, which spans a period of four hundred years and traces discoveries and the naming of succulents from the 1500s. Many of the books were part of the Richter gift, while others were donated or purchased. A conservation project to protect the books began in the early 1980s, funded in part by one of several IMS grants the Garden received. It included cataloguing, special custom boxes, and volunteer training.

"It's important to remember that these are not just things that are sitting there, they are used by the botanists because they are important," Jane Cole observed. "You try to protect them from casual use but you certainly make them available to people who need the information."

The Lois Porter Earle Herbarium

Jim Blakley didn't know it, but when he assembled the Garden's first herbarium sheets in 1950, he was putting together the initial pieces of the Garden's comprehensive research program. The twenty-two hundred specimens he assembled were augmented in 1953 by two important gifts—the George B. Hinton collection of northern and central Mexico specimens, donated by Mrs. Walter Douglas, and the Rose Collum collection of Arizona native plants, which included her field notes.

"The herbarium was originally in a closet," Blakley recalled. "We put in some shelves until we were able to get enough money for two herbarium cases. That was really great, because we could put things out where they should be. We just kept on going with little donations from members who knew how badly we needed things and we just grew and grew."

Blakley left the Garden in 1953 and the collection he assembled lay dormant until 1970, when J. Harry Lehr, a retired banker and self-taught botanist, joined the staff as volunteer curator of the herbarium. Lehr went to work restoring and reorganizing the collection and in 1972 the Lois Porter Earle Herbarium was opened under his care.

Due to the collecting, research, preservation techniques, and publications of Lehr and research botanist Dr. Howard Scott Gentry, the herbarium was designated as a National Resource Collection in 1976. Lehr continued to build the collection and in 1978 he published *A Catalogue of the Flora of Arizona*, a checklist of native Arizona plants. By the time he retired in 1984, the herbarium collection numbered in excess of twenty-six thousand specimens. Lehr also had documented eleven new species in Arizona.

Wendy Hodgson, who joined the staff in 1974 as an illustrator for Gentry, started collecting herbarium specimens in 1980 at Lehr's encouragement. She was named as Lehr's successor and over the next fifteen years nearly doubled the specimen count, resulting in a comprehensive collection of desert plants, Arizona flora, agaves, and cacti. With the assistance of Ted Anderson, Hodgson initiated a massive program in the 1990s to make herbarium specimens of the Garden's living collection.

A Garden prickly pear in bloom. Photo by Jennifer Johnston.

A living legacy

The collections, put together over the years by devoted staff and friends of the Garden, will serve as a legacy from people who thought far beyond their own lifetimes.

"These are long-living plants," Hodgson observed. "The people who collected these plants knew that. They didn't plant for themselves, but for future generations. That's really important, that people here were constantly thinking ahead."

Helia Bravo (left), a prominent Mexican botanist, visited W. Taylor Marshall (second from left) at the Garden in 1956. The couple to the right are Charles and Lillian Mieg.

CHAPTER EIGHT

THE RESEARCH PROGRAM

ROM THE ENTHUSIASTS who were part of Gustaf Starck's Sunday study group to the self-taught botanists and university trained professionals who led the way, the desire to generate knowledge about desert plants fueled the Desert Botanical Garden's development. Over the years, both amateur and professional researchers added immeasurably to the Garden's ability to carry out its mission and to the larger community's understanding of the desert ecosystem.

The early efforts

Although most of the activity in the Garden in the early years was centered around building the collection and facilities, research and learning was emphasized from the beginning. Directors made capturing data an important part of collecting trips. Staff produced articles concerning their studies of various plants.

Visiting professionals offered lectures and classes in their areas of expertise.

Enthusiastic self-trained members like Charlie Mieg often joined the staff and prominent visiting botanists on collecting trips. Many members contributed articles to the *Saguaroland Bulletin* on botanical topics and assisted the staff in collecting and disseminating information.

The Garden's first foray into formal research was Jim Blakley's herbarium work in 1950. Dr. James A. McCleary, a botany professor at Arizona State College at Tempe who served on the Garden's first Advisory Board, was named senior botanist on the staff in 1955. It was an unpaid position, which he occupied until 1960. McCleary made arrangements for his students to help with Garden projects in exchange for use of Garden research facilities.

Dr. Howard Scott

Gentry, the Garden's

first research botanist,

collected both living and

herbarium specimens

to support his study of the

agave. This photo was

taken by former director

Rodney Engard in

Coahuila, Mexico.

A landmark study

At Hubert Earle's request, John Rhuart, chairman of the Executive Board, wrote to Dr. Howard Scott Gentry in 1970 to offer him a position as the Garden's research botanist. The support for Gentry's position would come from a National Science Foundation grant to study agaves. Gentry agreed and joined the Garden staff in 1971.

Gentry, who became interested in agaves for their steroid potential, collected specimens in Mexico, Guatemala, Honduras, and Arizona, helping establish the Garden's collection of agaves as one of the most complete in the world. In 1982 he published *Agaves of Continental North America*, the culmination of more than thirty years of research and still considered the most complete work on the genus.

For the remainder of his tenure at the Garden, Gentry focused his research on economic botany, studying the economic potential of growing various desert plants as agricultural crops. He was delayed in publishing his agave monograph, held up when earlier research he had conducted on the economic potential of jojoba (*Simmondsia chinensis*) attracted the attention of the scientific community. Hundreds of scientists from around the world visited Gentry at the Garden to learn about jojoba. Gentry continued to concentrate on economic botany until his retirement in 1987.

A formal program of study

In Gentry, the Garden had a world-renowned researcher on board, but it wasn't until 1986 that it began to establish a broad-based research program.

Dr. Gary Nabhan joined the Garden

staff that year as assistant director for research. In concert with the Garden's membership in the Center for Plant Conservation, Nabhan expanded research efforts to include studies involving rare, threatened, and endangered plants. He focused attention on desert plant resource conservation and development, coordinating a number of projects between the Garden and Mexican researchers related to the ecology and conservation of borderland plants. Nabhan also initiated several studies on restoring degraded desert habitats.

Nabhan obtained many research grants to fund the Garden's expanding program, enabling him to hire more researchers. Rick DeLamater and Liz Slauson joined the staff in 1987 to assist with agave and rare plant research, while Dr. Alan Zimmerman became the cactus research botanist. Dr. Joe McAuliffe, Dr. Laura Jackson, and Pat Comus were hired in 1990 to work on the research involving degraded habitats.

During his tenure at the Garden, Nabhan earned both the prestigious Pew

Dr. Edward "Ted"

Anderson, an expert

on rare cacti and their

conservation, joined the

Garden staff in 1992 as

senior research botanist.

Memorial Trust and MacArthur Fellowship awards. His book, *Gathering the Desert*, received the John Burrows Medal for outstanding nature writing, and under his leadership the Garden received the first award for excellence in a research program from the American Association of Museums.

The Garden's research staff was publishing an average of twenty papers a year and by 1992, when Nabhan left the staff, he had developed a full-fledged research department. He was succeeded by McAuliffe.

Dr. Edward "Ted" Anderson, an expert on rare cacti and their conservation, joined the Garden in 1992 as senior research botanist and conducted numerous studies on rare and endangered cacti in the United States and Mexico. He also was integral in expanding the Garden's living and herbarium collections of cacti, donating hundreds of his own herbarium specimens.

As director of research, McAuliffe spearheaded the development of a mission statement for the department and mapped out a five-year plan. He also coordinated the collaboration of the research and education departments in developing educational material for the 1992-97 revamping of the Garden's interpretive trail system.

McAuliffe stepped down as director of research in October 1998 to devote time to several of his own research projects, remaining on the staff as plant ecologist. He was succeeded by Liz Slauson, who by that time had earned her doctorate.

Slauson had started as an intern in 1987 and became the Center for Plant Conservation botanist on the staff in 1988. She served as curator of the living collection and research botanist from 1990 until her appointment as director of research in 1998.

Researchers Wendy Hodgson, Dr. Liz Slauson and volunteer Amy Prince hiked the Grand Canyon National Park in search of rare agaves in September 1999.

Looking ahead

In 1998 the research department established a five-year plan to focus its future research efforts. The plan identified four areas where the department would expand its studies: systematics, ecology, conservation biology, and ethnobiology. Under the plan, the department would place major emphasis on developing a comprehensive research program for studies in the Mohave Desert and Sonoran Desert of Baja California.

From the first plant data collected in 1939 to the sophisticated and complex studies of 1999, research at the Desert Botanical Garden supported the Garden's mission to "study and disseminate knowledge" for sixty years. Building on those efforts with a cohesive plan, the research department was poised to make the Garden an even stronger leader in the study of desert plants.

The Garden's education department created exhibits for use in off-site venues.

CHAPTER NINE

REACHING OUT
TO THE COMMUNITY

ROM THE BEGINNING, the Desert Botanical Garden's link to the larger community essentially was its lifeline. The support of members and visitors was vital in the Garden's ability to fund its mission. To gain and maintain the flow of members and visitors to the Garden required not only an attractive physical setting, but interesting and worthwhile experiences as well.

An outdoor classroom

Visitors to the Garden in its early days followed trails of compacted dirt while using a numbered guidebook to identify the various plants along the way. The directors or visiting botanists offered lectures and slide shows in Webster Auditorium, and every so often a school group would arrive *en masse* for a tour.

Eventually, classes on weekdays and weekends were added to the schedule. Paths were blacktopped, the plant collec-

tion expanded, and field trips were formed. But it would be 1977—thirty-eight years after the Garden was founded—before the first steps were taken toward a formal education and interpretive program.

Until that time, there was little structure to providing guided tours of the Garden. Anyone wanting such a tour was taken through by the director, if he had time, or a member, if one was available. As the Garden grew, however, requests for tours grew with it. The newly established education department, headed by Sherry Couch Krummen, responded by creating a docent program, educating volunteers about the garden, and training them to conduct and manage tours.

The docent training program was extensively refined in the 1980s under the direction of Kathleen Socolofsky, teaching the volunteers how to illustrate concepts with plants and how to better address visitors' questions. More "hands-on" activities were incorporated into the tours and "touch carts" appeared along the

Garden docents were trained to illustrate concepts with plants through hands-on activities, enabling visitors to have a first-person experience with desert plants (photo by John Nemerovski).

trails, enabling visitors to have a first-person experience with desert plants.

"Kathleen really set the direction in this department," said Ruth Greenhouse, who became director of the Garden's Educational Services Department in 1998. "Her philosophy was to promote hands-on experiences and experiential learning, to provide opportunities for people not just to learn about these plants but to be able to touch, taste, and experience them."

While the early trails allowed visitors to walk through the Garden and view the plants, it wasn't until the 1980s that the Garden began to make significant enhancements that offered more involvement for visitors. The first was the opening in 1984 of the Rhuart Landscape Demonstration Garden, which later became part of the Center for Desert Living. The Rhuart exhibit showcased the benefits of landscaping with desert-adapted plants.

The second enhancement was the Plants and People of the Sonoran Desert Trail, which opened in 1988. It was the first trail to have educational signage and

activities for visitors and served as a model not only for future Garden projects, but for exhibits at many botanical gardens around the country as well.

"A lot was happening programmatically in the 1980s. Strategic planning and thinking was really beginning to move," Greenhouse said. "We built the trail, then Kathleen said we needed to do the same with the rest of the Garden. Robert Breunig sent her to Washington to get a National Science Foundation grant, which she did."

The grant Kathleen Socolofsky went after was the $634,000 NSF grant the Garden received in 1992. Along with funding from the cities of Phoenix and Scottsdale and private donations, the Garden used the grant to execute a comprehensive plan to improve the experience of its visitors. Improvements were made throughout the Garden for the comfort of visitors, with the most significant developments being concentrated in the trail system. The five-year $2-million project, completed in 1997, transformed the former maze of paths into an organized system of thematic trails, offering effective communication through carefully researched signage and interactive exhibits.

The Garden's early lectures and slide shows sowed the seeds for its public education program. Hampered for many years by a shortage of staff, volunteers, and funds, the Garden was limited in its ability to offer a variety of workshops and programs consistently. A growth in grants, donations, and manpower beginning in 1985 enabled the Garden to better reach out to the community.

School field trips benefited greatly from resources and training provided to teachers, trained volunteer field trip guides, a "Desert Detective" game, and puppet shows. A quarterly schedule of workshops offered a variety of new things for both children and adults to learn at the Garden.

A resource for the community

The people associated with the Desert Botanical Garden always have taken pride in its value to the community. In a speech in 1939, Gertrude Webster told her audience that "this botanical garden will enlighten the world as to the beauty and fascination of all desert growths." Those who came after Webster worked hard to make her vision a reality as the Garden became one of the Valley's top visitor attractions.

One important component of that work was to create events aimed at drawing community members to the Garden. The cactus shows from 1947 to 1987 routinely brought thousands of people to the Garden who might never have come otherwise. Special art and photography shows related to the desert also generated high attendance. Plant sales offered the public the opportunity to buy desert plants, many of which were not widely available anywhere else. The first Luminaria Night grew into a multi-night event, delighting thousands of visitors each year.

The spring wildflower season generated hundreds of calls to the Garden's Wildflower hotline each year, as did a Plant Questions hotline. The demand grew for the Garden's facilities for private parties and events, while many special events were held for members and the public. Art in Bloom, Dinner on the Desert, Music in the Garden and Jazz in the Garden concerts, family Garden nights, and other community-oriented activities filled the Garden's schedule each year.

The gift of time

While the community served as the Desert Botanical Garden's lifeline, the volunteers were its life force—the heart and soul of the Garden. The volunteer spirit began with the founding members of the Arizona Cactus and Native Flora Society, who traveled long distances over dusty roads to launch their beloved Garden. It continued through the many leaders who served on the Board of Trustees and its related committees, and the amateur botanists who accompanied the early directors on many a collecting trip.

Therese Marshall, wife of director W. Taylor Marshall, epitomized the dedication of the volunteers with her countless

The Education Department developed a "Desert Detective" game to give children an interactive experience at the Garden (photo by Pam McCarroll).

Sammy Saguaro greeted visitors and delighted children in the Garden as part of the education program.

hours of service behind "The Table" in Webster Auditorium. Lois Porter Earle, wife of director W. Hubert Earle, was responsible for so many unpaid jobs at the Garden that it took twenty or more volunteers to fill her place when she died.

The annual cactus show, the biggest event held at the Garden from 1947 to 1987, was staffed in large part by volunteers who put hundreds of hours into its planning and execution. Plant sales, photography and art exhibits, *Las Noches de Luminaria*, and many other public events all were made possible only through the help of volunteers.

Early volunteer efforts were loosely organized at best. Requests for specific events appeared in the bulletin and the few staff members knew which individuals they could call on for help whenever they needed it. Charlie Mieg's Cactomaniacs rounded up help for big events and Fred Beselt's Service Group brought in help after Lois Earle passed away. It was not until the docent program was organized under Rodney Engard's direction in 1977, however, that a planned volunteer effort got underway.

When Charles Huckins became director in 1979, he brought with him a model for a volunteer structure with a mission and objectives. He appointed Virginia Coltman, a staff member and former volunteer, to develop the "Friends of the Garden" as a part of the Membership Department. Coltman and volunteer Mary Bess Mulhollan organized the new group with the help of other staff and volunteers, recruited more than 160 "Friends," and put the organization to work in 1983 with Mulhollan as its first president. The group became "Volunteers in the Garden" in 1994 and by 1999 it included more than five hundred volunteers who worked in every area of the Garden.

Huckins instituted a formal recognition program for volunteers, presenting pins and certificates at the annual Volunteer Luncheon. He also began a program to document the number of volunteers and volunteer hours provided each year, which was necessary for the Garden to earn its accreditation from the American Association of Museums.

In 1987 the Garden's Education Department expanded its training curricula beyond the docent program to include volunteers interested in serving in other areas, such as horticulture or retail. By

1997 these volunteers were required to attend a fifty-hour volunteer core course before being placed in a specific area. Others, volunteers who were not required to complete the course, found opportunities in special events and clerical support for the Garden.

"The volunteer program has exploded from what it was years ago," said Frank Hennessey, who volunteered at the Garden from 1958 until 1998. "We had trouble getting twenty people together for an event. Now they've got them by the hundreds."

Hennessey's forty years of volunteer service were unusual, but not unique. The Garden was fortunate to have several volunteers who stuck with it year after year, helping ensure that it fulfilled its mission. By 1999 volunteers were contributing more

than sixty thousand hours of service, the equivalent of thirty full-time staff people.

A compelling attraction

In 1939 Gertrude Webster told the audience at the Garden's dedication that it was committed to becoming a "compelling attraction." Frank Hennessey, who saw many changes during his four decades at the Garden, observed that the staff and volunteers fulfilled that commitment beyond what any of the founders might have imagined.

"In its early days it didn't have near the attraction it has now. Things have really changed over the years, and for the better," Hennessey said. "I really love the garden and what's happened with it."

Archer House was renovated into offices in the mid-1908s. Pictured here in 1999, it housed the Education Department and the Volunteer Office.

CHAPTER TEN

PROPAGATING PLANS
FOR A NEW CENTURY

B Y 1999 the Desert Botanical Garden was poised to execute a strategic master plan. The Board of Trustees and the staff spent more than three years developing the plan, looking at its potential needs for the next twenty-five years.

The plan called for a new entrance, gift shop, and plant shop; a Desert Studies Center that would house the volunteers and education department and include facilities for the research, horticulture, and propagation departments; a new reception hall and gallery; and a new library and herbarium. Under the plan, the Garden also would renovate and expand existing exhibits, improve existing buildings, and add $1 million for endowment and additional support for working capital.

To fund this master plan, the Garden embarked on the first major capital development program in its sixty-year history. "Growing a Legacy for Generations" was launched in January 1999 and expected to raise $15.7 million by the end of the year.

"We're preparing for the future by improving our facilities and preparing for more trails and growth," Carolyn Polson O'Malley said. "The master plan and capital campaign will help us ensure that the Garden can fulfill its mission well into the next sixty years."

What would Gustaf Starck and Gertrude Divine Webster say about the way their beloved Desert Botanical Garden has developed over the years? One can only guess, but they would likely be pleased with the way the seeds they planted have grown and bloomed, staying true to the mission they outlined sixty years ago.

And what about the future? Perhaps Mary Irish, who joined the Garden as a volunteer in 1986 and later became director of public horticulture, summed it up best in the December 1998 *Sonoran Quarterly*:

Viewed from Webster Auditorium's roof in 1999, the Garden had grown and prospered remarkably in its first sixty years.

"Every planting is a little leap of faith, not only in the plant itself, but in the continuance of the Desert Botanical Garden and its long future. What might the Garden be like in sixty more years? The cardons will be pretty much full grown, seedling trees barely out of the greenhouse or the growing yard will be mature and well-settled matrons, countless agaves will have bloomed and died leaving equally countless pups, and the creeping devil will have run a few more miles around the base of Webster.

So while we celebrate the great past we have had and admire what the tiny sapling palo breas have become, we must constantly ask ourselves, *"What present are we preparing for those who will celebrate our one-hundred-twentieth-year anniversary?"* The planting of those seedlings is the kind of optimism that builds great gardens. It is certainly what got us to our sixtieth anniversary. In the end, using our past and present as a foundation for the future may be the greatest treasure that any public garden can offer its community."